Field Guide To:

USEFUL TREES AND SHRUBS IN DRYLAND PERMACULTURE

By

Dr. Keller Horton

Published by Kellerman Consulting
Printed in the United States of America

Book Cover and Interior Design by Dan Fowler
Editing by Keller Horton and Dan Fowler

Library of Congress
Cataloging-in-Publication Data

Horton, Keller
Field Guide To: Useful Trees and Shrubs In Dryland Permaculture
1st Edition
ISBN-13: 978-0692936412
ISBN-10: 0692936416

Table of Contents

Introduction

The reason I wrote this book is simple. There are so many of us in permaculture who still have trouble remembering which leaves, seeds, seed pods, flowers, and general tree shapes belong to some of the important Nitrogen Fixers and other trees that are often used in dryland permaculture. I also included several trees that most permaculture folks like to use due to their usefulness in producing foods, medicines, and for attracting beneficial insects.

For those of you who don't yet know, a Nitrogen Fixer is a tree or plant that has little nodules on its roots that house a very wonderful type of bacteria. This bacteria actually converts the Nitrogen from our air into deep, ground-penetrating fertilizer, and the leaves of these plants and trees are also an excellent source of high nitrogen fertilizer! Imagine what these wonderful plants can do for any soil that has become completely dead and useless. Their use is one of the cornerstones of permaculture. When used correctly, the Nitrogen Fixers and other trees in this book will provide nutrients and shade for a plethora of fruits, nuts and vegetables on your land.

Another important factor I address in this book is whether or not your goats, sheep, or even pigs, can eat these plants in order to bring down the cost of feeding your animals. Furthermore, I wanted to inform the readers, in the quickest way possible, about the cold/heat hardiness, shade provision, and a few pertinent facts about their horticulture, food production and other uses in a farm setting.

One last important item. I wanted this book to be eye catching and beautiful. I want your friends to pick it up and ask you questions about permaculture.

When you design a permaculture site I want you to remember that you are creating a gorgeous food forest that pays homage to the millions of forests that our earth has created all on its own. It's our job to creatively follow the earth's design principles rather than fight against them. Thank you and enjoy.

About The Author

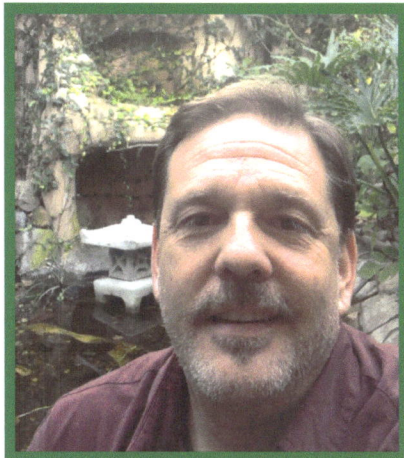

Doctor Keller Horton is not a medical doctor. He comes from a background of thirty five years in education and curriculum design. During that time he also completed his Master's Degree specializing in Environmental Law and Regional Design Methodologies for Development in Lower Income Countries.

Dr. Horton has over 50 years of gardening experience. His lessons began early in life with his parents and both sets of grandparents on farms in West Texas. He caught the Permaculture, "bug" while doing You Tube research on aquaculture greenhouses. While watching a video on aquaculture, a mysterious thumbnail video title appeared on the right side of the screen. The title simply read, "Greening the Desert". The temptation to click on that video was unbearable. When he watched the video, an entire new world of creating self-sustaining food forests was revealed.

Twenty-seven books and 100 videos later he decided to enroll in the fantastic permaculture course offered by the Permaculture Research Institute in Jordan. Yes, Jordan, in the Middle East where the Jordan river separates the nation of Israel from the Hashemite Kingdom of Jordan. The course site was located right within one of the hottest, driest, most rocky, and inhospitable spots you could ever imagine on Planet Earth.

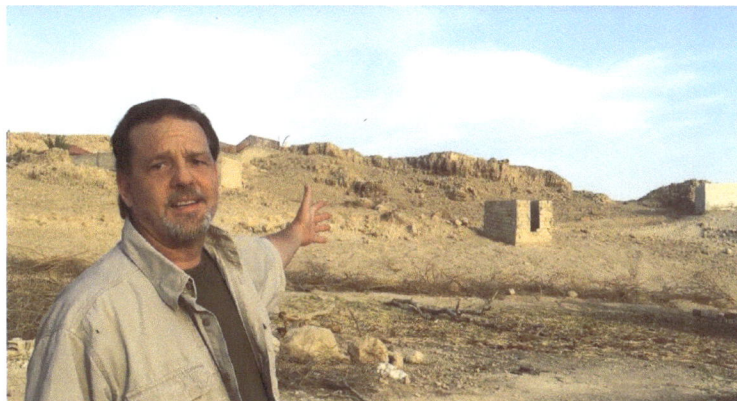

Yet, there Mr. Geoff Lawton and his wife Nadia had designed and created an absolutely GORGEOUS Permaculture Food Forest. It is covered in shade and filled with over 100 types of Nuts, fruits, and vegetables! You must see it to believe it.

There Dr. Horton met both Mr. and Mrs. Lawton, the directors of the Permaculture Research Institute, and he completed the life changing Permaculture Design Course located right there in the driest, rockiest part of the Jordan Valley.

Currently, Dr. Horton resides in California where he maintains his Permaculture Design practice, participates in environmental activism, and is beginning the conversion of a 40 acre parcel in the Mojave Desert into a Permaculture Food Forest.

Acknowledgements

First and foremost, I must thank my parents for insisting that every home we ever lived in was surrounded with trees for fruit, nuts and mulch, and all had space for a garden. Growing food was an integral part of my childhood.

Second, I am thankful to Geoff and Nadia Lawton for continually and tirelessly spreading information about permaculture and making it crystal clear that this is not just an earth-friendly way to design a sustainable garden and farm. It is a great opportunity to make a darn good living whilst healing this planet. Many farmers increase their bottom line by up to 400% using permaculture design principles!

Furthermore, I am deeply grateful to the entire Al-Bashir family of Amman, Jordan; who have committed to work with me in converting every farm that they own over to permaculture design principles. Their willingness to experiment and their encouragement have been both an inspiration and a blessing. I am truly fortunate to have begun to know the Al-Bashirs as clients and continue to know them as life long friends.

Finally, I could not have finished this book without the unending support, creative giftedness, and technical expertise of my dear friend, Dan Fowler. His encouragement is worth more than any earthly treasures that can be imagined

Just A Quick Reminder:

The species in this book are those that you can use in permaculture to help turn your dryland "dirt" into a shady, mulch covered, rich soil that is filled with life.

We also want to produce TONS of food. Fruits and nuts that will grow well in your dry land or desert area can include:

- Dates
- Olives
- Figs
- Mulberries
- Prickly Pear
- Banana
- Yucca
- Citrus
- Avocados
- Walnuts
- Mukau nut from East Africa
- Paw Paw
- Yellow Passion Fruit
- Almonds
- Pomegranates
- And so many more!

And remember, all of your trees will need to be <u>drip irrigated for about 5 years</u> to get a healthy head start.

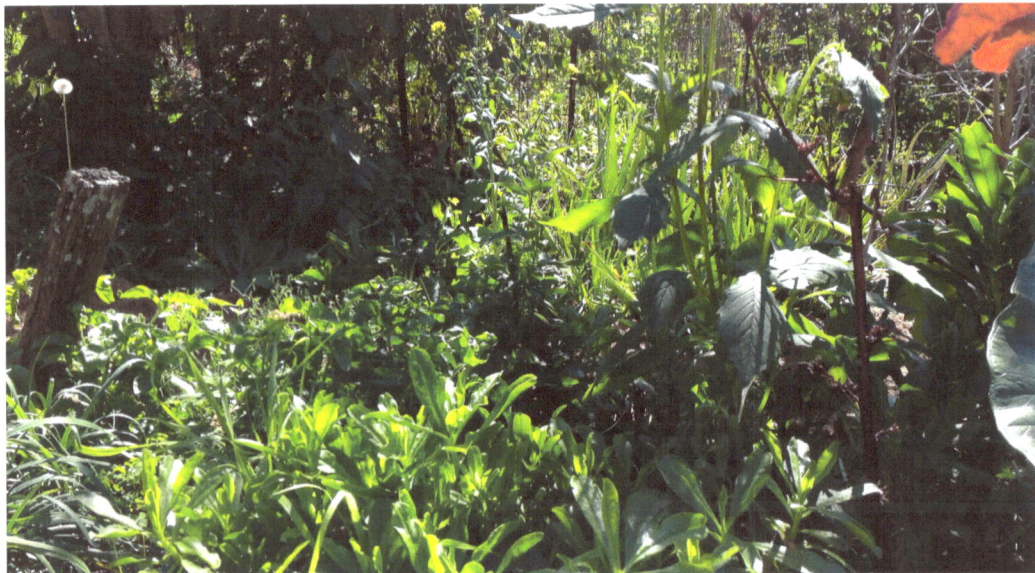

Date palm overstory, with hundreds of fruits and vegetables in this 2,000 year old Moroccan permaculture food forest.

GROUP ONE:
THE NITROGEN FIXERS

Notes
(Gum Arabic Tree)

Acacia arabica
(Gum Arabic Tree)

a)   Height- to 20 meters
b)   Good fodder? Yes (overdose is unhealthy)
c)   Good Nitrogen fixer? Yes
d)   Cold tolerance -Yes
e)   Salt tolerant? Yes
f)   Wide shade canopy? Medium

Notes: Used for Gum Arabic, medicines, toothbrushes, timber, and tanning leather, windbreaks, thorny fencing.

Notes
(Golden Wreath Wattle)

Acacia saligna
(Golden Wreath Wattle)

a) Height - 9 meters
b) Fodder? Yes
c) Nitrogen fixer? Yes
d) Cold tolerance - To 4°C
e) Salt tolerance? Yes, slight
f) Wide canopy? No
g) Good mulch? Yes

Notes: Used for timber, fuel wood, charcoal, tannin, thorny fencing. Boil seeds before planting.

Notes
(Acacia victoriae)

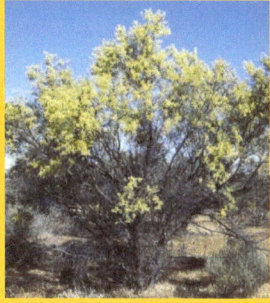

Acacia victoriae

a) Height- 5 meters
b) Good fodder? Yes
c) Nitrogen fixer? Yes
d) Cold tolerance - Thrives down to 5°C, ok down to - 5°C
e) Salt tolerant? Yes
f) Wide canopy? No
g) Good mulch? Yes

Notes: Slow grower. Used for charcoal and fuel wood. Seeds used to make baking flour. Good thick windbreak. Boil seeds before planting.

Notes
(Casuarina)

Casuarina glauca
(Casuarina)

a) Height- 10 -15 meters
b) Fodder? Yes
c) Nitrogen fixer? Yes
d) Cold tolerance - To -5°C
e) Salt tolerant? Yes
f) Wide shade canopy? No
g) Good mulch? Yes

Notes: Reduces alkali in soil. Good windbreak.

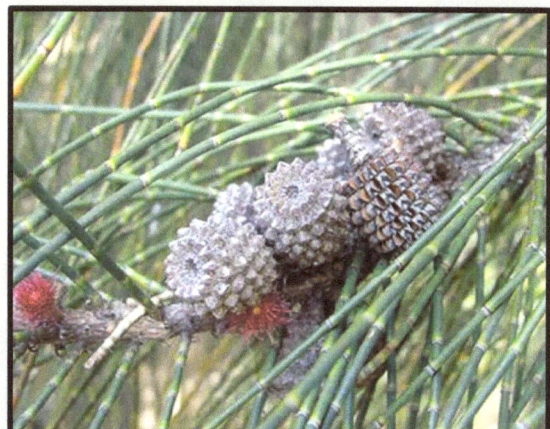

Notes
(Carob Tree)

Ceratonia siliqua
(Carob Tree)

a) Height- Up to 15 meters
b) Good fodder? Yes
c) Nitrogen fixer? Yes
d) Cold tolerance - Down to 0° C
e) Salt tolerant? Yes, slight
f) Wide canopy? Yes, to 15M meters
g) Good mulch? Yes

Notes: Bees love it. Reduces erosion. Roots go down 20M. Soak seeds in warm water for 3 days prior to planting. Grow seedlings in pipes to encourage tap root. Stake baby trees at 20 cm high. Seeds good food like chocolate, can be made into a good liquor.

Notes
(Mesquite Tree)

Prosopis glandulosa
(Mesquite Tree)

a) Height? Fast growing. Often to 10 meters. Velvet Mesquite up to 18 meters
b) Fodder? Yes
c) Nitrogen fixer? Yes
d) Cold tolerance - Yes. Down to -15°C
e) Salt tolerance? Yes.
f) Wide canopy? Up to 12 meters if trained
g) Good mulch? Yes.

Notes: Roots can grow 50 meters down to find water. Pod with beans can be ground into flour to make sweet bread. Scar seeds before planting and refrigerate for 2 weeks before storing in hot water for 2 days to encourage sprouting. Ambient temperature must be between 27°C and 29°C for seeds to germinate (80° - 85°F).

Notes
(Black Locust)

Robina pseudoacacia
(Black Locust)

a) Height- 25 meters
b) Good fodder? No, toxic
c) Nitrogen fixer? Yes, slight
d) Cold tolerant? Yes, snow
e) Salt tolerant? Yes, medium
f) Wide canopy? Yes, 15 meters
g) Good mulch? Yes

Notes: Seeds useful food. Fruit is like vanilla. Flowers make delicious tea. Several medicinal uses. Good erosion control. Good for bees. Bark used as yellow dye.

Notes
(Scarlet Wisteria)

Sesbania grandiflora
(Scarlet Wisteria)

a)   Height- Up to 15 meters (Sesbania sesban only up to 7 meters)
b)   Good fodder? Yes. Must cut and carry. Must mix in with other fodder.
c)   Nitrogen fixer? Yes
d)   Cold tolerance - Frost sensitive. Dies in prolonged cold periods.
e)   Salt tolerant? Yes
f)   Wide canopy? No, good for planting near plants that need only slight shade.
g)   Good mulch? Yes

Notes: Fast growing. Not wind tolerant. Leaves, fruits, and flowers are edible for humans. Hundreds of medicinal uses. Grows well with peppers, coffee, tea and cocoa.

2 mm

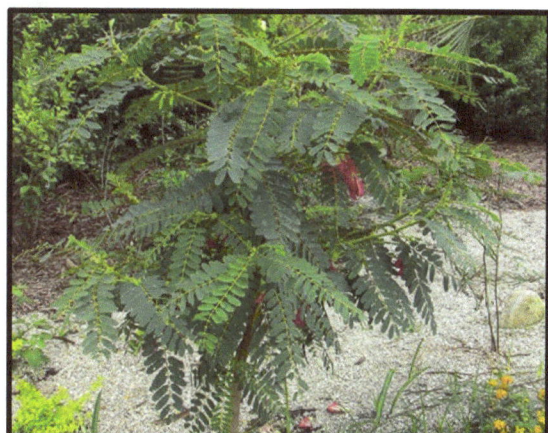

More Notes

GROUP TWO:
ANIMAL FODDER & MORE

Notes
(Old Man Salt Bush)

Atriplex nummularia
(Old Man Salt Bush)

a) Height – up to 3 meters
b) Good fodder? Yes. Must be mixed in with other fodders.
c) Nitrogen fixer? No
d) Cold tolerant? Yes, Down to 1°C.
e) Salt tolerant? Yes, very salt tolerant.
f) Large canopy? No, it is a shrub.
g) Good mulch? No, leaves are too salty.

Notes: Easy to propagate with seeds or shoots. Needs good drainage but tolerates occasional flooding. Grows back quickly after cutting or grazing. Red legged earth mites try to attack it but are prevented my good mulch cover and diverse plantings.

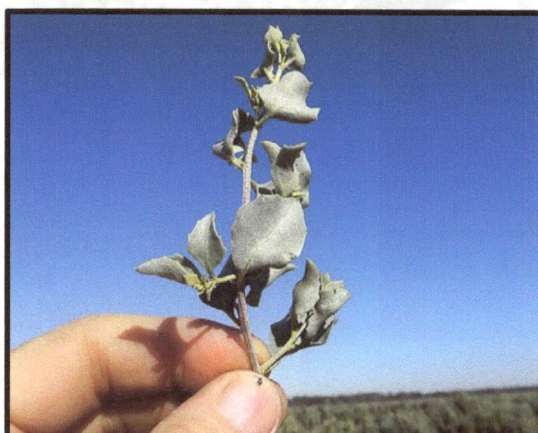

Notes
(Blue Jacaranda)

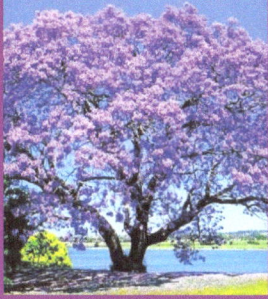

Jacaranda mimosifolia
(Blue Jacaranda)

a) Height - to 25 meters
b) Good fodder? Yes
c) Nitrogen fixer? No
d) Cold tolerance - Down to 0°C
e) Salt tolerance? Yes, slight
f) Large canopy? Yes
g) Good for mulch? Yes

Notes: Good windbreak. Good timber. Large blue flower clusters.

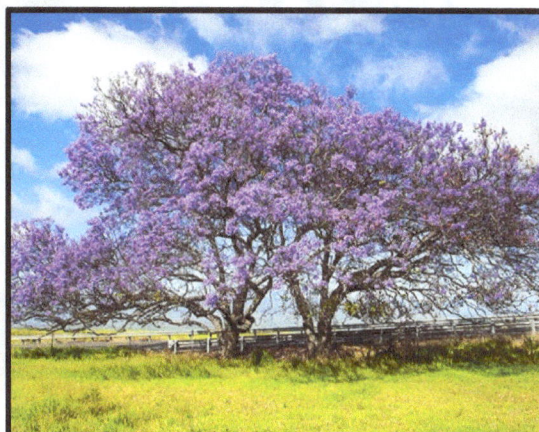

Notes
(Jerusalem Thorn)

Parkinsonia aculeata
(Jerusalem Thorn)

a) Height - up to 6 meters
b) Good fodder? Yes
c) Nitrogen fixer? No
d) Cold tolerant? Yes
e) Salt tolerant? Yes
f) Wide canopy? Yes, if trained.
g) Good mulch? Yes

Notes: Bright yellow flowers. Good thorny fence.

Notes
(Pickleweed, Sea Asparagus)

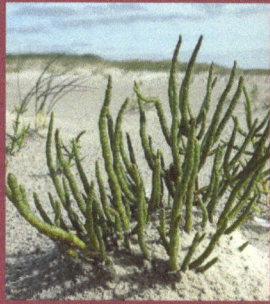

Salicornia procumbens
(Pickleweed, Sea Asparagus)

a) Height - less than 1 meter
b) Good fodder? Yes
c) Nitrogen fixer? No
d) Cold tolerant? Yes
e) Salt tolerant? Yes, very. Can be watered with sea water.
f) Wide canopy? No, it's a ground cover
g) Good mulch? Yes

Notes: Salicornia oil is good for cooking. The leaves can be cooked and eaten as a vegetable or added to salads. The dried biomass can be used to make wood boards, paper and fuel briquettes.

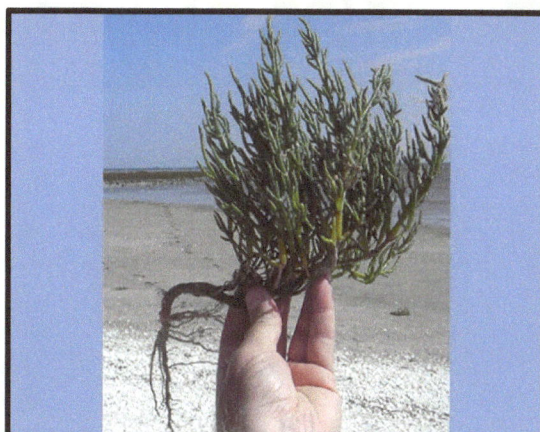

Notes
(Mediterranean Saltwort)

Salsola vermiculata
(Mediterranean Saltwort)

a) Height – 1 meter short.
b) Good fodder? Yes
c) Nitrogen fixer? No
d) Cold tolerance - Down to 1°C
e) Salt tolerant? Yes, very.
f) Wide canopy? No, it's a shrub.
g) Good mulch? Yes

Notes: Really good goat and sheep fodder.

Notes
(Salt Cedar)

Tamarix jordanis
(Salt Cedar)

a) Height - up to 15 meter
b) Good fodder? Yes
c) Nitrogen fixer? No
d) Cold tolerance - Below 0°C
e) Salt tolerant? Yes, very.
f) Wide canopy? To 5 meters with training
g) Good mulch? Yes

Notes: Deep taproot. Good windbreak. Grows well in Jordan Valley. Good erosion control. Used for medicines. Bark used for eczema. Good firewood. Bark and flowers used for tanning leather.

More Notes

GROUP THREE:
FOOD, MEDICINES & SALVES

Notes
(Yarrow)

Achillea millefolium
(Yarrow)

a) Height - Lacy, thin, fern like flower, 1 to 2 meters
b) Good fodder? Yes, must mix with others.
c) Nitrogen fixer? No
d) Cold tolerant? Yes, tolerates snow.
e) Salt tolerant? Yes
f) Wide canopy? No, it's a thin lacy flower.
g) Good mulch? Yes

Notes: Deep roots bring nutrients up to topsoil. Needs sun, dies in shade. Prevents erosion. Attracts pollinators including butterflies and ladybugs. Edible for people. Several medicinal uses. Yarrow oil will kill mosquito larvae when added to water.

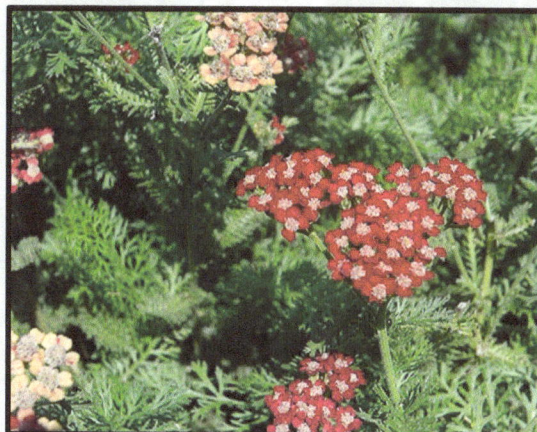

Notes
(Greek Strawberry Tree)

Arbutus andrachne
(Greek Strawberry Tree)

a) Height - Normally 6 meters, up to 12 meters
b) Good fodder? Yes
c) Nitrogen fixer? No
d) Cold tolerance- down to -15°C
e) Salt tolerant? Yes, moderate.
f) Wide canopy? Yes, with training.
g) Good mulch? Yes

Notes: Needs sun. Slow grower. Soak seeds for 5 days before planting. Attracts bees, butterflies, and hummingbirds.

Notes
(Neem Tree)

Azadirachta indica
(Neem Tree)

a) Height - up to 25 meters
b) Good fodder? Yes, mixed in as supplement.
c) Nitrogen fixer? No
d) Cold tolerance - down to 1°C
e) Salt tolerant? Yes
f) Large canopy? Yes, very wide
g) Good mulch? Yes, excellent mulch.

Notes: Wind resistant. Leaves are healthy vegetable for humans, high in vitamins. Used to make medicines, shampoos, creams. Twigs can be used as a toothbrush. Good erosion control. Anti-malarial. Good quality timber. Oil used to rid pests from plants.

Notes
(Bay Leaf) (Ghar)

Laurus nobilis
(Bay Leaf) (Ghar)

a) Height - to 7.5 meters
b) Good fodder? No
c) Nitrogen fixer? No
d) Cold tolerance - to -5°C
e) Salt tolerant? Yes, slight. And alkali tolerant.
f) Wide canopy? No
g) Good mulch? Yes

Notes: Fragrant flavor for soups and stews. Repels Meal Moth, roaches, mice and silverfish. Leaves are healthy antioxidant. Used as anti-parasite medicine et al.

Notes
(African Drumstick Tree)

Moringa oleifera
(African Drumstick Tree)

a) Height - 12 meters
b) Good fodder? Yes
c) Nitrogen fixer? No
d) Cold tolerant? Mild frost ok
e) Salt tolerant? Yes, slight
f) Wide canopy? Yes, 12 meters
g) Good mulch? Yes

Notes: Leaves are healthy vegetable. Oil is great for skin and hair, antibiotic, anti-itch, good mosquito repellant, good bath oil. Grows very fast.

Notes
(Christ's Thorn Jujube)

Ziziphus spina-cristi
(Christ's Thorn Jujube)

a) Height - Normally 6-9 meters, up to 20 meters
b) Good fodder? Yes, only ok.
c) Nitrogen fixer? No
d) Cold tolerance - Down to -23°C
e) Salt tolerant? Yes
f) Wide canopy? Yes, if trained
g) Good mulch? Yes

Notes: A type of Magnolia. Deep taproot. Seeds only germinate after going through digestive tract of mammals. Good thorny hedge and windbreak. Type of Jujube. Fruit paste can be baked into bread. Used to make medicines, a narcotic and shampoo. Wood is good for carving. Grows well with Millet.

Extra Resources & Reading Materials

For more information on Permaculture visit these websites:

The Permaculture Research Institute- http://permaculturenews.org
www.geofflawtononline.com/videos

AND READ THE FOLLOWING BOOKS

I.   Sowing Seeds in the Desert by Masanobu Fukuoka

II. The One Straw Revolution by Masanobu Fukuoka

III. Sepp Holzer's Permaculture: A Practical Guide to Small-Scale, Integrative Farming and Gardening by Sepp Holzer

IV. Rainwater Harvesting for Drylands and Beyond by Brad Lancaster

V.   Introduction to Permaculture by Bill Mollison

VI. Permaculture: A Designers' Manual by Bill Mollison

www.ingramcontent.com/pod-product-compliance
Lightning Source LLC
Chambersburg PA
CBHW052048190326
41521CB00002BA/142